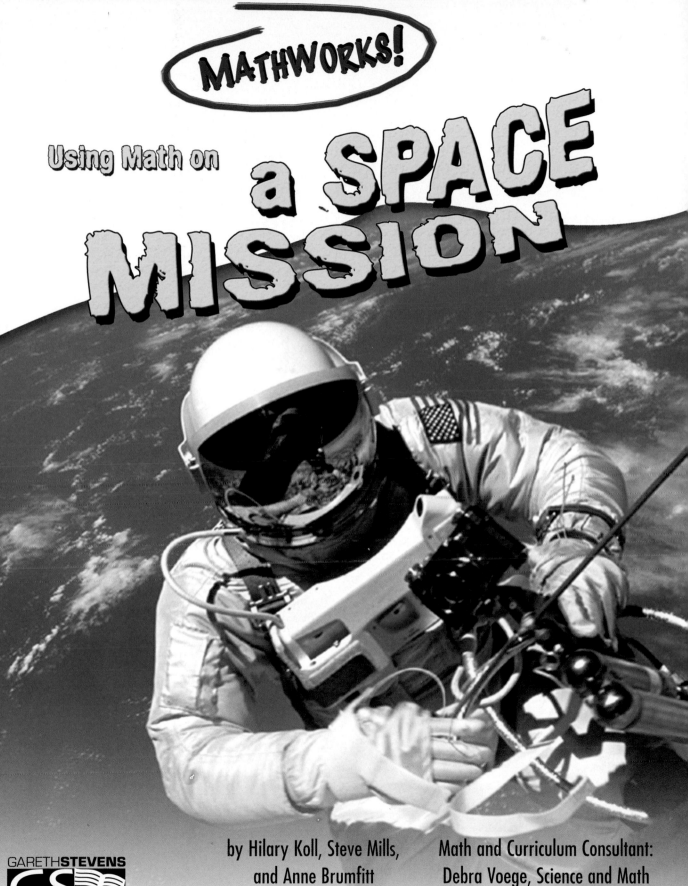

MATHWORKS!

Using Math on a SPACE MISSION

by Hilary Koll, Steve Mills, and Anne Brumfitt

Math and Curriculum Consultant: Debra Voege, Science and Math Curriculum Resource Teacher

GARETH**STEVENS**
GS
PUBLISHING
A Member of the WRC Media Family of Companies

Please visit our web site at: **www.garethstevens.com**
For a free color catalog describing Gareth Stevens Publishing's list of high-quality books and multimedia programs, call 1-800-542-2595 (USA) or 1-800-387-3178 (Canada). Gareth Stevens Publishing's fax: (414) 332-3567.

Library of Congress Cataloging-in-Publication Data available upon request from publisher.
Fax (414) 336-0157 for the attention of the Publishing Records Department.

ISBN-10: 0-8368-6763-7 — ISBN-13: 978-0-8368-6763-3 (lib. bdg.)
ISBN-10: 0-8368-6770-X — ISBN-13: 978-0-8368-6770-1 (softcover)

This North American edition first published in 2007 by
Gareth Stevens Publishing
A Member of the WRC Media Family of Companies
330 West Olive Street, Suite 100
Milwaukee, Wisconsin 53212

This U.S. edition copyright © 2007 by Gareth Stevens, Inc.
Original edition copyright © 2006 by ticktock Entertainment Ltd.
First published in Great Britain in 2006 by ticktock Media Ltd.,
Unit 2, Orchard Business Centre, North Farm Road,
Tunbridge Wells, Kent, TN2 3XF, United Kingdom.

Technical Consultant: Anne Brumfitt is the education consultant to the European Space Agency Directorate of Science in Noordwijk, Netherlands, and has extensive background as a classroom teacher and special education consultant.

Gareth Stevens Editors: Dorothy L. Gibbs and Monica Rausch
Gareth Stevens Art Direction: Tammy West

Photo credits (t=top, b=bottom, c=center, l=left, r=right)
NASA: 6br (Goddard Space Flight Center), 6-7t (J. Bell–Cornell U. and M. Wolff–SSI), 6-7b, 20-21, 22-23, 24-25 (Pat Rawlings), 26-27; ESA: 8-9 (D. Ducros), 12-13, 14-15 (AOES Medialab), 16-17, 18-19 (D. Ducros); ESA/Starsem: 10-11; ESA/DLR/FU Berlin: 13tr (G. Neukum).

Printed in the United States of America

1 2 3 4 5 6 7 8 9 10 09 08 07 06

CONTENTS

HAVE FUN WITH MATH

How to Use This Book

Math is important in the daily lives of people everywhere. We use math when we play games, ride bicycles, or go shopping, and everyone uses math at work. Imagine you are an astronaut, and you are going to travel into space. You may not realize it, but astronauts use math to plan and carry out their space explorations. In this book, you will be able to try lots of exciting math activities, using real-life data and facts about space missions. If you can work with numbers, measurements, shapes, charts, and diagrams, then you could BE PART OF A SPACE MISSION.

How does it feel to travel in space?

Put on your space suit and find out what it takes to plan a mission into space.

Math Activities

The mission clipboards have math activities for you to try. Get your pencil, ruler, and notebook (for figuring out problems and listing answers).

THE INTERNATIONAL SPACE STATION

The information from Concordia can help you determine how humans might deal with cold, dark conditions for long periods of time, but it cannot help you find out how humans might handle being in space. You need to know whether people find living in weightless conditions stressful as well as how to prevent damage to human bones and muscles. The International Space Station (ISS) was set up outside Earth's atmosphere to collect information for long missions in space. Teams of researchers live on the ISS for months at a time. You decide to go to the ISS to find out for yourself what life is like in space.

Mission File

In the DATA BOX on page 19, you will see information about the weight of a person on different planets and on the Sun and the Moon.

If a person weighs 100 pounds on Earth, how much will that person weight on
1) Jupiter?
2) Neptune?
3) Saturn and Uranus?
4) Venus?
5) Mars?
6) Mercury?
7) Pluto?
8) the Sun?
9) the Moon?

SPACE TOILET FACTS

• Space toilets do not use water. Because of weightlessness in space, astronauts must fasten themselves to the toilet seat. A lever operates a powerful fan, and a suction hole slides open. The air stream carries the waste neatly away.
• Some ISS crew members find it difficult to get used to using the toilet in space. Besides getting used to the device itself, they have to get used to the unusual fact that their bowels actually float inside their bodies — like the rest of their internal organs and, of course, everything else on board the ISS.

18

NEED HELP?

- If you are not sure how to do some of the math problems, turn to pages 28 and 29, where you will find lots of tips to help get you started.

- Turn to pages 30 and 31 to check your answers. (Try all the activities and challenges before you look at the answers.)

- Turn to page 32 for definitions of some words and terms used in this book.

You will find lots of amazing details about space missions in FACT boxes that look like this.

ISS FACTS
- The ISS travels around Earth at about 250 miles above our heads.
- The ISS orbits Earth at a speed of 17,400 miles per hour, which is about 4.8 miles every second.
- The ISS takes only 1.5 hours to orbit Earth.

Solar panels on the International Space Station provide the station with energy.

DATA BOX — Gravity

Because each planet's gravity is different, a person's weight will change if he or she goes to different planets.

Planet	Approximate Weight
Jupiter	2.6 times the weight on Earth
Neptune	1.4 times the weight on Earth
Saturn	1.1 times the weight on Earth
Uranus	1.1 times the weight on Earth
Venus	0.9 times the weight on Earth
Mars	0.4 times the weight on Earth
Mercury	0.3 times the weight on Earth
Pluto	0.1 times the weight on Earth

On the Sun, a person's weight is about 280 times his or her weight on Earth. On the Moon, a person's weight is about 0.2 times his or her weight on Earth.

Math Facts and Data

To complete some of the math activities, you will need information from a DATA BOX that looks like this.

Math Challenge

The ISS orbits Earth at 17,400 miles per hour. To help you understand how fast this speed is, you can compare it to the speeds of other moving objects.

A person walks at a speed of 5 miles per hour. This speed is also 8,800 yards per hour, 146.7 yards per minute, and 2.4 yards per second. Now try answering these questions. Round your answers to one decimal place. You will need a calculator.

1) A person runs at a speed of 10 miles per hour. What is this speed in
   a) yards per hour?   b) yards per minute?   c) yards per second?

2) A car on a highway travels at a speed of 60 miles per hour. What is this speed in
   a) yards per hour?   b) yards per minute?   c) yards per second?

3) An express train travels at a speed of 150 miles per hour. What is this speed in
   a) yards per hour?   b) yards per minute?   c) yards per second?

4) An airplane travels at a speed of 500 miles per hour. What is this speed in
   a) yards per hour?   b) yards per minute?   c) yards per second?

5) The ISS travels at a speed of 17,400 miles per hour. What is this speed in
   a) yards per hour?   b) yards per minute?   c) yards per second?

19

Math Challenge

Blue boxes, like this one, have extra math questions to challenge you. Give them a try!

THE AMAZING PLANET MARS

Scientists believe Mars is more like Earth than any of the other planets, and, for a long time, people have wondered whether humans could ever live there. Further from the Sun than Earth, Mars is a cold, dry place. It is smaller than Earth, but its surface is fascinating, with towering mountains and deep craters. The highest mountain on Mars, called Olympus Mons, is actually a volcano. Due to billions of years of erosion by Martian winds, the planet is covered with dust. There is so much dust on Mars that winds have blown it into dunes at the planet's north pole. In other places, winds have blown the dust away, leaving bare rock. The facts and data on this page will give you more information about this amazing planet.

Mars

Mission File

The DATA BOX below compares Mars and Earth.
Use the information to help you answer these questions.

1) How many more
   a) minutes are there in one Earth day on Mars than on Earth?
   b) Earth days are there in one year on Mars than on Earth?

2) How much smaller is Mars' than Earth's
   a) equatorial radius?   c) core radius?
   b) polar radius?

3) About how many times higher is Mars' highest mountain, Olympus Mons, than Earth's highest mountain, Mount Everest?
   a) 2 times   b) 2½ times   c) 3 times

Polar radius

Equatorial radius

Core radius

DATA BOX Mars and Earth

Name of Planet	Mars	Earth
Equatorial radius	2,111 miles	3,963 miles
Polar radius	2,097 miles	3,950 miles
Core radius	1,057 miles	2,165 miles
Number of Earth days in one year	687	365
Number of hours in one Earth day	24 hours, 37 minutes	24 hours
Height of highest mountain	82,021 feet	29,035 feet

MARS FACT

Mars is sometimes called the Red Planet because of its color. Oxidized iron in Mars' soil and rocks makes the planet look red.

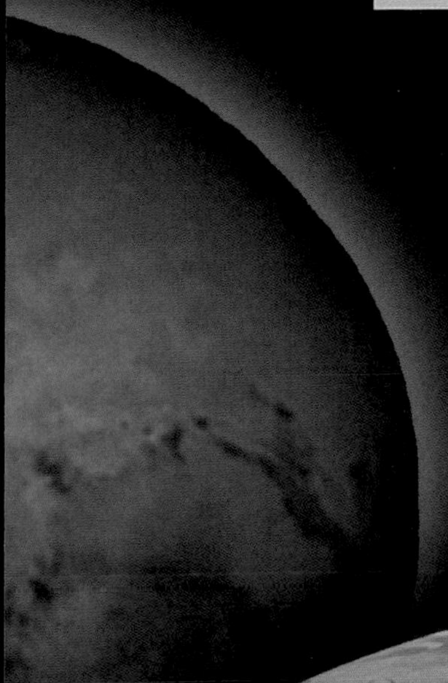

Scientists hope to send astronauts to Mars by the year 2030.

DATA BOX

Distance from the Sun

All of the planets in our solar system revolve around the Sun.

This chart shows average distances, rounded to the nearest million miles, from the Sun.

Earth	93 million miles
Jupiter	484 million miles
Mars	142 million miles
Mercury	36 million miles
Neptune	2,795 million miles
Pluto	3,667 million miles
Saturn	887 million miles
Uranus	1,784 million miles
Venus	67 million miles

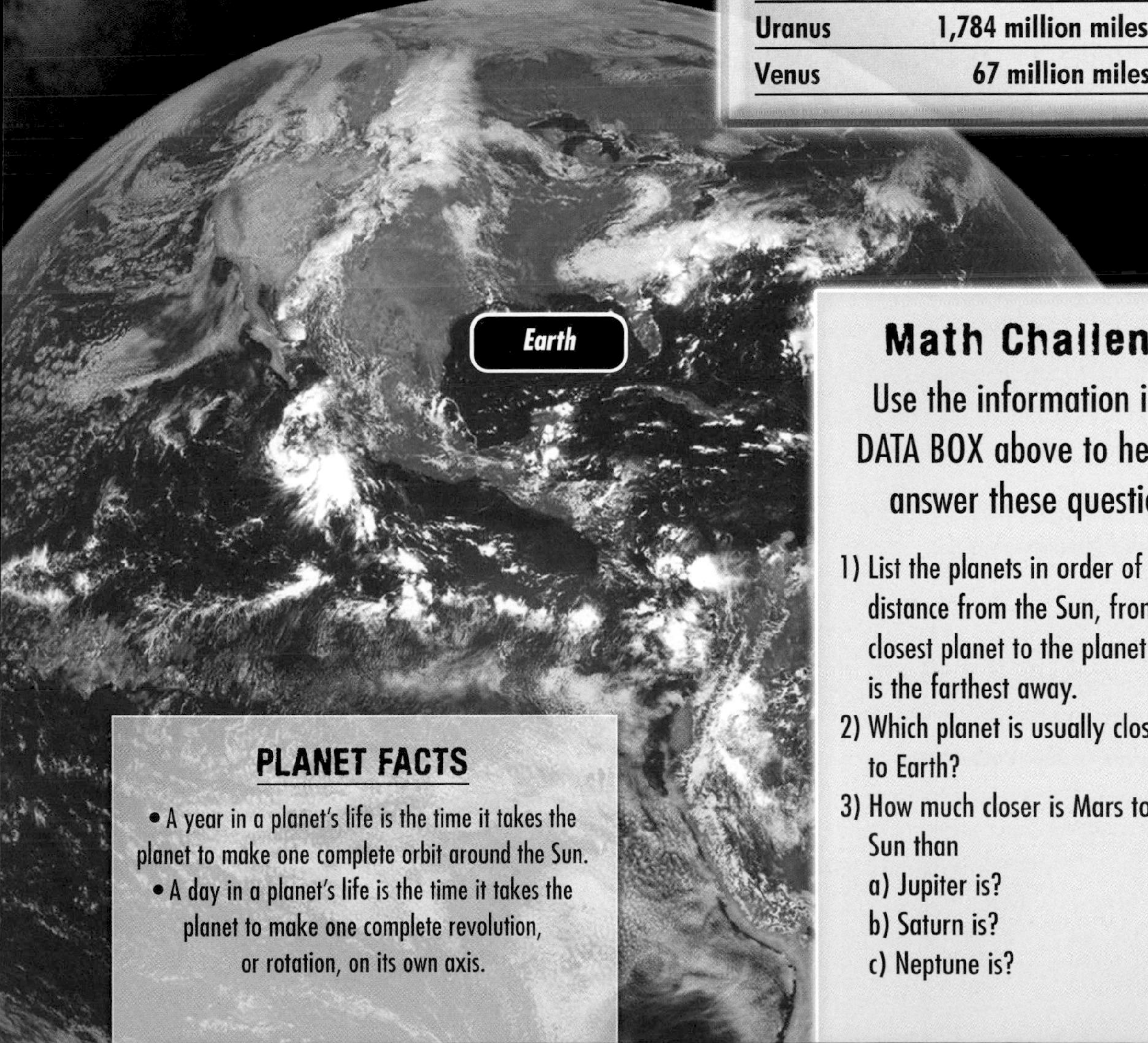

Earth

Math Challenge

Use the information in the DATA BOX above to help you answer these questions.

1) List the planets in order of average distance from the Sun, from the closest planet to the planet that is the farthest away.
2) Which planet is usually closest to Earth?
3) How much closer is Mars to the Sun than
   a) Jupiter is?
   b) Saturn is?
   c) Neptune is?

PLANET FACTS

• A year in a planet's life is the time it takes the planet to make one complete orbit around the Sun.
• A day in a planet's life is the time it takes the planet to make one complete revolution, or rotation, on its own axis.

SPACE PROBE

The space agency you work for eventually wants to send a manned spacecraft to Mars. Before they can send this spacecraft, however, they need to learn much more about the planet. The space agency decides to send a robotic space probe to Mars. The probe, called Mars Express, will not have humans on board. It will orbit, or revolve around, Mars to collect data without risking any human lives. Mars Express is designed to be low cost, quick, and efficient. Its tasks are to look for water below the surface of Mars, to analyze the atmosphere, and to look at the geology of the planet. All of this information is important to know before sending people on a mission to Mars.

Mission File

In the DATA BOX on page 9, you will see information on the orbits of Mars and Earth. You need to launch the probe when the two planets are close together so the probe travels the least amount of distance.

Use the DATA BOX to answer these questions.

1) You decide to launch the probe exactly one hundred days before the date when Mars and Earth are the closest. What will the date of launch be if you want to land a probe on Mars in
   a) 2010?
   b) 2012?

2) Look at your answers to 1a and 1b. If the journey takes exactly six months, on what dates will the probe reach Mars?

SOLAR PANEL FACT

Mars Express has solar panels that use sunlight to provide the probe with energy. The panels charge batteries that help the probe keep working when it travels to the side of Mars away from the Sun, where sunlight does not reach the solar panels. When the probe is attached to a rocket for launching, the solar panels are folded up. After the probe separates from the rocket, the panels are opened, one at a time, and turned to face the Sun. The solar panels have an area of 122.93 square feet. They can generate 650 watts of electricity.

DATA BOX # The Orbits of Mars and Earth

The circular path an object takes as it moves around another object is called an orbit. In our solar system, all planets move in an orbit around the Sun.

Earth takes about 365 days to orbit the Sun. Mars takes 687 days. Because Earth and Mars take different lengths of time to orbit the Sun, the two planets are sometimes close to each other and sometimes far apart. The diagrams below show how the distance between Earth and Mars changes.

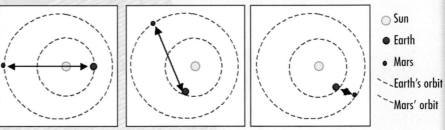

○ Sun
● Earth
• Mars
- - Earth's orbit
- - Mars' orbit

The Mars Express mission must be launched at a time when Mars is at its closest point to Earth. In 2003, Mars was very close to Earth. It was only about 34 million miles away. Sometimes, Mars can be as far away as 250 million miles away. Earth and Mars will take just over two years and one month (about 765 days) to come close to each other again.

Below are some of the dates when Mars will be closest to Earth.

December	January	March	April
23	**27**	**4**	**10**
2007	2010	2012	2014

Math Challenge

The area of the solar panels on Mars Express is 122.93 square feet. The number line below shows decimal points between 122.5 and 123. Can you find 122.93 on this line?

TAKE OFF!

he day has arrived to launch Mars Express. A countdown begins twelve hours before the launch. Many systems and equipment need to be tested and monitored before liftoff. A launch countdown is planned carefully so all the people involved in the mission have time to check everything needed for a successful launch. The countdown also ensures that tests are carried out in the right order. You are asked to help with the final countdown by making sure that each test or task happens on time. As the end of the countdown approaches, everyone waits anxiously to see if the rocket takes off successfully.

Mission File

In the DATA BOX on page 11, you will see the launch countdown, beginning twelve hours before the launch. Use the information to answer these questions.

1) If the launch time is supposed to be at exactly 17:45 (5:45 p.m.), at what time would each of these stages take place?

a) Stage 1   f) Stage 6
b) Stage 2   g) Stage 7
c) Stage 3   h) Stage 8
d) Stage 4   i) Stage 9
e) Stage 5

2) You have been given the job of counting down the steps on several tasks. You need to count backwards in different sized steps. Can you figure out what the next three numbers in each of these sequences will be?

a) 100, 90, 80, 70, 60, 50, 40, . . .
b) 18, 16, 14, 12, 10, . . .
c) 42, 37, 32, 27, 22, 17, . . .
d) 24, 21, 18, 15, 12, . . .
e) 29, 25, 21, 17, 13, . . .

DATA BOX Launch Countdown

Stage	Time	Event
1	-12h 00m 00s	Start final countdown
2	- 7h 30m 00s	Check electrical systems
3	- 3h 20m 00s	Chill down the main engine
4	- 1h 10m 00s	Check connections between launcher and command systems
5	-7m 00s	Give "all systems go" report to allow start of sequence
6	-4m 00s	Pressurize tanks for flight
7	-1m 00s	Switch to onboard power mode
8	-04s	Onboard systems take over
9	-03s	Unlock guidance systems to flight mode
LAUNCH!	00s	Ignite main engine

Mars Express was launched with the Soyuz rocket.

Math Challenge

In the Rocket Fact box below, you will see how fast a rocket has to travel to go into orbit around Earth.

If the rocket continued to travel at this speed, how far would the rocket travel in

1) 2 seconds?
2) 10 seconds?
3) 30 seconds?
4) 1 minute?
5) 5 minutes?
6) 10 minutes?
7) 30 minutes?
8) 1 hour?

ROCKET FACT

A rocket must use tremendous power to reach what is known as "escape velocity." Escape velocity is the speed at which a rocket must be traveling to get away from Earth's gravity. Gravity is the force that pulls objects toward the center of a planet. To go into orbit around Earth, a launched rocket has to travel at a speed of 5 miles per second. This speed is one hundred times faster than most racing cars can travel.

MISSION CONTROL

The launch was successful, and you are now at Mission Control for Mars Express. Mission Control is the place where every part of the probe's journey is monitored and, if necessary, adjusted. The control room is full of computer screens showing different types of information about the probe. Mars Express continuously sends millions of bits of data back to this ground station. The data show unique details of the planet Mars. Countless evaluations are made from these data so scientists can unveil the secrets of Mars and its atmosphere. You need to be able to read the data and find out what it means.

Mission File

These dials and scales appear on the computer screens at Mission Control.

What is the number each red arrow is pointing to?

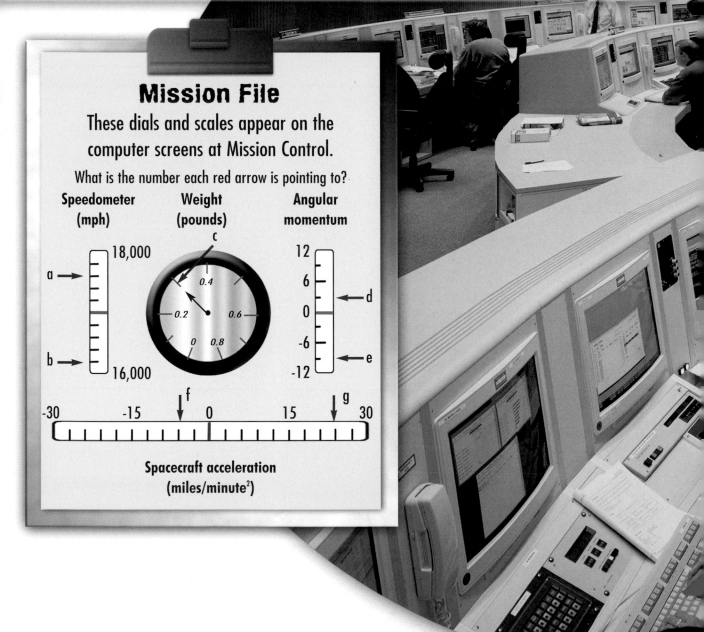

Speedometer (mph)

a →
b →

18,000

16,000

Weight (pounds)

c

0.4
0.2   0.6
0   0.8

Angular momentum

12
6
0
-6
-12

← d
← e

-30   -15   0   15   30

f   g

Spacecraft acceleration
(miles/minute²)

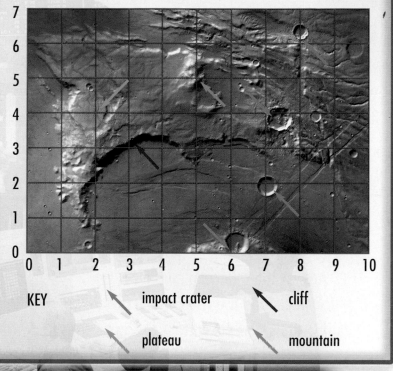

DATA BOX Ground Control

Mars Express sent this photo of the surface of Mars. Each square on the grid is about 9 miles across.

KEY   impact crater   cliff

plateau   mountain

Math Challenge

Everyone at Mission Control must analyze and record the information from Mars Express carefully and correctly. Use the grid in the DATA BOX above to help you answer these questions about Mars' surface.

Can you find the missing coordinates to match each of these features in the photo?

1) Large impact craters can be found at (6,0) and (7,) and near (7,) and (8,).

2) A large circular plateau can be found near (,4).

3) The tallest cliff in this photo can be found near (3,).

4) The highest mountain in this photo can be found near (,5).

13

Six months after the launch, Mars Express safely reaches its destination. The probe remains in orbit around Mars, while a lander is sent down to the surface. You use scientific instruments to direct the probe and lander to collect information. You need the probe to take photos and to create maps of the planet's surface. You also must try to measure the temperature of Mars at different places on the surface. You have so many questions you need to answer. Does Mars' atmosphere have oxygen for humans to breathe? What is the planet's geology like? How great is the force of gravity on Mars?

Mission File

Answer the following questions to understand the rise and fall of temperatures on the surface of Mars.

1) At the south pole of Mars, a temperature of −112 °F rose by 117 °F. What is the new temperature?

2) In summer, a daytime temperature of 46 °F fell by 90 °F. What is the new temperature?

3) Find the new temperatures if
   a) a temperature of 28 °F fell by 50 °F.
   b) a temperature of 27 °F rose by 11 °F, then rose again by 12 °F.
   c) a temperature of 5 °F fell by 27 °F, then rose by 50 °F.
   d) a temperature of −103 °F, rose by 42 °F, then rose by 48 °F, and then fell by 29 °F.

An artist shows how a lander on Mars might look as it begins to explore the planet.

Temperatures on Mars

Mars Express is collecting lots of data about the temperature on the surface of Mars. From these measurements, you can find the highest and lowest temperatures of different places on the surface.

Maximum temperature: 80 °F (daytime at the equator in summer)
Minimum temperature: −207 °F (nighttime at Mars' north pole in winter)

You discover there is a difference between the temperature on the ground and the temperature 3 feet above the ground. Sometimes, the difference in temperature can be as much as 60 °F.

The graph below shows two sets of temperatures on Mars, one taken 3 feet above the ground (the pink line) and one taken 10 inches above the ground (the blue line).

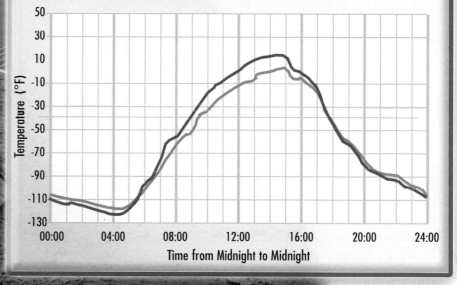

Math Challenge

Look carefully at the line graph in the DATA BOX above.
Which of these statements are true and which statements are false?

1) The temperature at 12:00 noon is warmer at 3 feet above the ground than at 10 inches above the ground.
2) The temperature at 18:00 is the same at 3 feet above the ground as it is at 10 inches above the ground.
3) The temperature at 04:00 is colder at 3 feet above the ground than at 10 inches above the ground.
4) The highest recorded temperature is about 10 °F.
5) The lowest recorded temperature is about −125 °C.
6) The lowest temperature was recorded at about 02:00.

CONCORDIA

Because you are receiving so much information from the probe about how cold it is on Mars, you feel that more research is needed. You want to find out how humans might cope, for long periods of time, in conditions that are very cold and very dark. You decide to send a research team to Concordia, which is a space research center in Antarctica. In winter, Concordia has less than one hour of sunlight a day for several months, and temperatures fall to as low as −121 °F. You need to monitor how people survive and stay healthy in such cold, dark conditions to determine how they might survive and stay healthy on Mars.

Concordia is a permanent research station in Antarctica.

Mission File

In the DATA BOX on page 17, you will see a table showing temperatures and wind speeds at the Concordia research station. Use the table to help you answer these questions.

1) In which month or months is the
   a) coldest average air temperature?
   b) warmest average air temperature?
   c) highest maximum air temperature?
   d) lowest minimum air temperature?
   e) fastest average wind speed?
   f) slowest average wind speed?
   g) fastest maximum wind speed?

2) What is the difference between the maximum and minimum air temperatures in
   a) April?
   b) August?
   c) June?

3) How many more miles per hour (mph) faster is the average wind speed in
   a) April than in August?
   b) June than in March?
   c) October than in February?

CONCORDIA FACTS

- Concordia is very remote and isolated. Its isolation makes Concordia ideal for testing equipment and procedures for future work on other planets and on the Moon.
- Concordia is an excellent place for understanding the behavior of small groups of people living for long periods of time in confined spaces, such as in a spacecraft, space station, or planet or moon base.
- About sixteen people work at the station in winter, and about thirty-two work there in summer. The workers include nine scientists, four technicians, a chief chef, an assistant cook, and a doctor.

BUILDINGS FACT

The winter research center at Concordia has three buildings linked by walkways.
One building has sleeping quarters, a communications room, laboratories, and a hospital.
A second building has a workshop, a kitchen, and a restaurant. A third building has a
wastewater treatment plant, a boiler room, a generator, and another workshop.

DATA BOX Concordia Statistics

The table below shows temperatures and wind speeds during one year at the Concordia
research station in Antarctica. The orange rows in the table are summer months, and the
blue rows are winter months, when there is less than one hour of sunlight per day.

Month	Average air temp (°F)	Maximum air temp (°F)	Minimum air temp (°F)	Average wind speed (mph)	Maximum wind speed (mph)
January	-19.5	3.2	-49	6.7	18
February	-43.2	1.4	-72	4	13.4
March	-68	-21	-92	5.4	20
April	-77	-47.2	-95	5.4	20
May	-85	-24	-112	6.5	26.8
June	-77.8	-44	-108.2	6.9	24.6
July	-77.2	-36	-100	5.8	22.4
August	-90	-50.8	-112	4.3	15.6
September	-70	-21	-104	5.6	22.4
October	-59	-24.5	-88.8	6.3	26.8
November	-36.8	7.9	-79	5.1	22.4
December	-24.2	1.4	-48	5	18

Math Challenge

You can find the mean of the average air temperatures for the
whole year by adding all of the average air temperatures for
each month, then dividing by the number of months.

Find the mean average air temperature for the year. Round your answer to the nearest
tenth (one decimal place).

THE INTERNATIONAL SPACE STATION

The information from Concordia can help you determine how humans might deal with cold, dark conditions for long periods of time, but it cannot help you find out how humans might handle being in space. You need to know whether people find living in weightless conditions stressful as well as how to prevent damage to human bones and muscles. The International Space Station (ISS) was set up outside Earth's atmosphere to collect information for long missions in space. Teams of researchers live on the ISS for months at a time. You decide to go to the ISS to find out for yourself what life is like in space.

Mission File

In the DATA BOX on page 19, you will see information about the weight of a person on different planets and on the Sun and the Moon.

If a person weighs 100 pounds on Earth, how much will that person weigh on
1) Jupiter?
2) Neptune?
3) Saturn and Uranus?
4) Venus?
5) Mars?
6) Mercury?
7) Pluto?
8) the Sun?
9) the Moon?

SPACE TOILET FACTS

• Space toilets do not use water. Because of weightlessness in space, astronauts must fasten themselves to the toilet seat. A lever operates a powerful fan, and a suction hole slides open. The air stream carries the waste neatly away.

• Some ISS crew members find it difficult to get used to using the toilet in space. Besides getting used to the device itself, they have to get used to the unusual fact that their bowels actually float inside their bodies — like the rest of their internal organs and, of course, everything else on board the ISS.

- The ISS travels around Earth at about 250 miles above our heads.
- The ISS orbits Earth at a speed of 17,400 miles per hour, which is about 4.8 miles every second.
- The ISS takes only 1.5 hours to orbit Earth.

Solar panels on the International Space Station provide the station with energy.

DATA BOX   Gravity

Because each planet's gravity is different, a person's weight will change if he or she goes to different planets.

Planet	Approximate Weight
Jupiter	2.6 times the weight on Earth
Neptune	1.4 times the weight on Earth
Saturn	1.1 times the weight on Earth
Uranus	1.1 times the weight on Earth
Venus	0.9 times the weight on Earth
Mars	0.4 times the weight on Earth
Mercury	0.3 times the weight on Earth
Pluto	0.1 times the weight on Earth

On the Sun, a person's weight is about 280 times his or her weight on Earth. On the Moon, a person's weight is about 0.2 times his or her weight on Earth.

Math Challenge

The ISS orbits Earth at 17,400 miles per hour. To help you understand how fast this speed is, you can compare it to the speeds of other moving objects.

A person walks at a speed of 5 miles per hour. This speed is also 8,800 yards per hour, 146.7 yards per minute, and 2.4 yards per second. Now try answering these questions. Round your answers to one decimal place. You will need a calculator.

1) A person runs at a speed of 10 miles per hour. What is this speed in
   a) yards per hour?   b) yards per minute?   c) yards per second?

2) A car on a highway travels at a speed of 60 miles per hour. What is this speed in
   a) yards per hour?   b) yards per minute?   c) yards per second?

3) An express train travels at a speed of 150 miles per hour. What is this speed in
   a) yards per hour?   b) yards per minute?   c) yards per second?

4) An airplane travels at a speed of 500 miles per hour. What is this speed in
   a) yards per hour?   b) yards per minute?   c) yards per second?

5) The ISS travels at a speed of 17,400 miles per hour. What is this speed in
   a) yards per hour?   b) yards per minute?   c) yards per second?

DAILY ROUTINE ON THE ISS

A s you settle in aboard the International Space Station, your first duty is to maintain the station. Besides cleaning and vacuuming dust from surfaces and the air, you also clean filters and Install and upgrade computer software. You spend a lot of time exercising, too. Muscles deteriorate in little or no gravity. Forcing your muscles to work keeps them from becoming weak. Finally, you must monitor scientific experiments. Some of the experiments require working with chemicals or performing delicate crystal-growing tests. You have plenty to do during your stay on the ISS.

Mission File

The DATA BOX on page 21 shows how astronauts spend their time. Use the information to answer these questions.

1) How many hours a week does an astronaut usually work?

2) What percentage of time on a shift is not spent on maintenance, exercise, or research?

3) Write each of these percentages as a fraction in its simplest form.
   a) 20%　　　　b) 30%　　　　c) 25%

Astronaut Sergei Krikalev exercises on the Treadmill Vibration Isolation System (TVIS).

ISS FACTS

What would happen if you stepped outside the ISS without a space suit? DON'T LEAVE!

• You would lose consciousness in as little as 15 seconds because there is no oxygen outside the ISS. Death would follow quickly.

• Because there is no air pressure to keep your blood and body fluids in a liquid state, the fluids would "boil," then freeze, before they evaporated completely. This entire process would take only 30 seconds to 1 minute.

• Your skin, heart, and other organs in your body would expand because of the boiling fluids.

• You would face extreme temperature changes, from 250 °F in sunlight to −150 °F in darkness.

• You would be exposed to various types of radiation (cosmic rays) or electrically charged particles coming from the Sun (solar wind).

• You probably would be hit by bits of dust or rock moving at high speeds or by orbiting debris from spacecraft or satellites.

DATA BOX

A Day on the ISS

Astronauts on the ISS work regular hours when possible — just as they do on Earth. They try to work eight-hour shifts during the week and half days on Saturday. Sunday is a rest day (a day off).

This pie chart shows how an astronaut might spend time during an eight-hour shift.

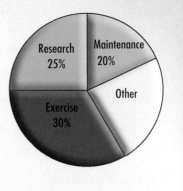

Research 25%
Maintenance 20%
Other
Exercise 30%

French astronaut Claudie Haigneré spent a week on the International Space Station.

Math Challenge

The ISS orbits Earth every 1½ hours. After astronauts on the ISS see the Sun rise, they have 45 minutes of daylight. Then the Sun sets, and the astronauts have 45 minutes of darkness, when Earth is between the ISS and the Sun.

Answer the following questions using the 24-hour military clock.

1) The sun rises at 04:40. List the times of the next three sun rises you will see.

2) The sun sets at 09:55. List the times of the next three sunsets you will see.

FOOD AND WATER ON THE ISS

Y̲ou will be staying on the ISS for up to six months. Do you know what you will eat and drink there? The ISS does not have a refrigerator so all of your food must be canned, dehydrated, or packaged so it does not need to be kept cold. Because the food comes in containers that can be thrown away, no dishwashing is necessary. Space station crew members eat three meals a day. Drinks on the ISS come in powdered form, and crew members have only lukewarm, warm, and hot water with which to make their drinks. No cold water is available. Crew members usually eat breakfast and dinner together.

Mission File

In the DATA BOX on page 23, you will see a list of one day's meals on the ISS.

The menu aboard the ISS repeats every eight days for the length of the mission, which is sixty days. Continue the next numbers in each sequence to show the days on which the astronaut will eat the same foods.

1) Menu 1 is served on days 1, 9, 17, 25, . . .
2) Menu 3 is served on days 3, 11, 19, 27, . . .
3) Menu 6 is served on days 6, 14, 22, 30, . . .
4) Menu 8 is served on days 8, 16, 24, 32, . . .

Even food floats aboard the ISS.

WATER FACTS

Besides air, water is the most important element aboard the ISS. Water on the ISS originally was brought from Earth, and the water has to be used carefully. Astronauts cannot take long, luxurious showers. Most astronauts just take sponge baths. The water recovery system on the ISS will collect, recycle, and distribute water from these places:

- the sink
- the shower
- space suit wastewater
- heating and cooling systems
- the space shuttle's fuel cells

- urine (from both the astronauts and the laboratory animals on board the ISS)
- cabin air (Astronauts and laboratory animals exhale moisture into the air.)

Before leaving Earth, all crew members taste every food item and score each item according to how well they like it. The scores are used to plan menus for the crew. The menus are checked by nutritionists before the food is packed and delivered to the space station. Astronauts need the same number of calories in space as they need on Earth. A small woman, for example, requires about 1,900 calories a day, while a large man requires about 3,200 calories. Most of the vitamins and minerals the astronauts need in space are also the same ones they need on Earth.

DATA BOX The ISS Menu

Below is an example of an ISS astronaut's daily menu.

To understand the menu, you need to know these words.
Rehydratable describes food that needs water added to it before it can be eaten.
Thermostabilized describes food that has been preserved by heat, which destroys any microorganisms.
Intermediate moisture describes food that is packaged with some, but not all, moisture removed.

MENU 1

Breakfast
Cottage cheese with nuts (rehydratable)
Oatmeal with peaches (rehydratable)
Plum-cherry dessert (intermediate moisture)
Coffee with sugar

Midday Meal
Grilled chicken (thermostabilized)
Rice with butter (thermostabilized)
Creamed spinach (rehydratable)
Pineapple (thermostabilized)
Grapefruit drink

Evening Meal
Chicken fajitas (thermostabilized)
Tortillas
Southwestern corn (thermostabilized)
Apples with spice (thermostabilized)
Brownie
Peach-apricot drink

Snack
Dried pears (intermediate moisture)
Nuts
Orange-pineapple drink

Math Challenge

The menu in the DATA BOX above shows the food an astronaut on the ISS ate on October 19.

If the menu is repeated every eight days, find the next four dates when the astronaut will eat the same food.

MISSION TO MARS

You have gathered data from the Mars Express probe, the Concordia research station, and the ISS. Now you can start thinking about a manned mission to Mars! Because the trip will probably take at least two years, you need to plan the mission carefully. The spacecraft cannot carry enough food, water, and air for each person for two years, so you will have to find ways to recycle air and water and to grow food. The astronauts on the mission will be the first people to ever set foot on another planet. Sending them 37 million miles into space – and bringing them back safely – will be one of the most impressive achievements in human history.

Mission File

It is the year 2030, and astronauts are just about to land on Mars. The newspapers are full of stories about what might happen when the astronauts land. You rate each event by how likely it is to happen.

1) The astronauts find that Mars is made of red cheese.   impossible

2) The spacecraft lands in the correct area.   likely

3) The astronauts discover that Mars once had water.   even chance

4) The landing damages the spacecraft.   unlikely

5) The astronauts refuse to explore the planet.   very unlikely

6) The astronauts are the first people on Mars.   certain

7) The astronauts discover that Mars once had life.   unlikely

8) The astronauts bring back rocks never seen on Earth.   very likely

Probabilities can be shown on a probability scale, from impossible to certain. Where should each event above go on this scale?

impossible   even chance   certain

After a two-year journey to Mars, astronauts will be excited to finally explore the planet.

Mars Sample Return

The National Aeronautics and Space Administration (NASA) is planning several missions to Mars. One of the most exciting missions is the Mars Sample Return project, due to be launched in the next ten to fifteen years. The goal of this mission is to bring back rocks from Mars so scientists can study the rocks in detail.

It is the year 2030, and your Mars probe has collected rocks from four different sites, or places, on Mars. When the probe returns to Earth, the rocks are distributed to space agencies around the world for study. This table shows how many rocks were collected from each site on Mars and where the rocks were sent.

Agency	Site 1	Site 2	Site 3	Site 4
NASA	3	5	7	2
European Space Agency (ESA)	0	2	3	1
Japan Aerospace Exploration Agency (JAXA)	4	1	3	2
Russian Federal Space Agency (RKA)	3	4	4	1
Indian Space Research Organization (ISRO)	1	1	1	0

Math Challenge

Use the table in the DATA BOX above to help you answer these questions.

1) How many rocks is NASA studying?
2) How many rocks were collected from Site 2?
3) From which site were the most rocks collected?
4) How many more rocks is Russia studying than India?
5) How many rocks were collected in total?

BECOMING AN ASTRONAUT

t will happen someday, but it will be many years before scientists have enough information to safely send people to Mars. Maybe you will be the first human to visit another planet. If you want to take part in a mission to Mars, the first step is to become an astronaut. Becoming an astronaut is not easy! There are very few places for training and a lot of competition for the positions. Many apply, but very few people become astronauts. If you are lucky, you may one day have the unique opportunity to see Earth from space and to experience the feeling of weightlessness. Some astronauts get so used to being weightless that, back on Earth, they simply let go of objects and are surprised to see them fall crashing to the ground.

Mission File

People have predicted that humans will walk on Mars by the year 2030. Most astronauts are between 28 and 40 years old.

Find how old each of these people will be in the year 2030 and whether they will be the right age in 2030 to be astronauts.

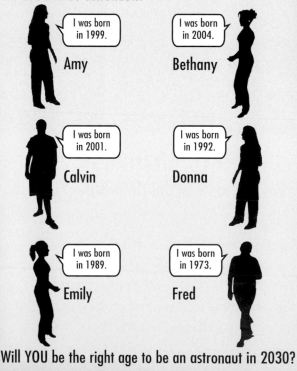

I was born in 1999.

Amy

I was born in 2004.

Bethany

I was born in 2001.

Calvin

I was born in 1992.

Donna

I was born in 1989.

Emily

I was born in 1973.

Fred

Will YOU be the right age to be an astronaut in 2030?

SPACE SUIT FACTS

To protect the astronaut from harm, a space suit must
- have a pressurized atmosphere.
- provide oxygen and get rid of carbon dioxide.
- maintain a comfortable temperature, even during strenuous work and movement into and out of sunlit areas.
- protect against from micrometeoroids and radiation.
- let the astronaut see clearly.
- let the astronaut move around easily.
- let the astronaut communicate with others (controllers on the ground, other astronauts, etc.)

ASTRONAUT FACTS

It is not be easy to become an astronaut, but if you wanted to do something easy, you would not want to be an astronaut! Here is what you will need to qualify.

1) You must have the right background and knowledge. Many astronauts began as pilots in military air forces. Other astronauts have scientific backgrounds in physics, engineering, or medicine.

2) Because being an astronaut is demanding work, you must be healthy and physically fit.

3) You must speak English well. Astronauts come from all over the world, but they all speak English so they can communicate with each other.

4) On a space station, you will live and work in a small area, so it is important to get along well with other people.

5) You must be determined and motivated to become an astronaut. You will spend hundreds, or even thousands, of hours training.

Math Challenge

In the year 1961, Yuri Gagarin became the first human in space. If the first human on Mars lands there in the year 2030, how much time will have passed between these two events?

Give your answer   1) in years   2) in months

MATH TIPS

PAGES 6-7

Mission File

When subtracting numbers with many digits, make sure you line up the digits — the tens with the tens, the hundreds with the hundreds, and so on.

```
  2,870
-   228
  2,642
```

PAGES 8-9

Mission File

You might find these calendar pages useful. Also, remember that 98 days is equal to 14 weeks.

SEPTEMBER

					1	2
3	4	5	6	7	8	9
10	11	12	13	14	15	16
17	18	19	20	21	22	23
24	25	26	27	28	29	30

OCTOBER

1	2	3	4	5	6	7
8	9	10	11	12	13	14
15	16	17	18	19	20	21
22	23	24	25	26	27	28
29	30	31				

NOVEMBER

		1	2	3	4	
5	6	7	8	9	10	11
12	13	14	15	16	17	18
19	20	21	22	23	24	25
26	27	28	29	30		

DECEMBER

					1	2
3	4	5	6	7	8	9
10	11	12	13	14	15	16
17	18	19	20	21	22	23
24	25	26	27	28	29	30
31						

JANUARY

1	2	3	4	5	6	
7	8	9	10	11	12	13
14	15	16	17	18	19	20
21	22	23	24	25	26	27
28	29	30	31			

FEBRUARY

				1	2	3
4	5	6	7	8	9	10
11	12	13	14	15	16	17
18	19	20	21	22	23	24
25	26	27	28	29		

MARCH

					1	2
3	4	5	6	7	8	9
10	11	12	13	14	15	16
17	18	19	20	21	22	23
24	25	26	27	28	29	30
31						

This calendar shows a leap year, which 2012 will be.

PAGES 10-11

Mission File

When continuing a sequence, find the difference between the numbers next to each other and look for a pattern.

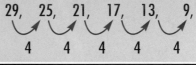

29, 25, 21, 17, 13, 9,

4   4   4   4   4

PAGES 12-13

Mission File

Follow these steps when reading scales on measuring instruments.

Step 1: Choose two numbers on the scale that are next to each other, and find the difference between them.

Step 2: Count how many intervals, or equal spaces, are between the numbers.

Step 3: Divide Step 1 by Step 2 to find the value of each interval.

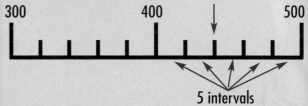

5 intervals

Each interval on the scale above has a value of 20, so the arrow is pointing to 440. $(400 + 20 + 20 = 440)$

Math Challenge

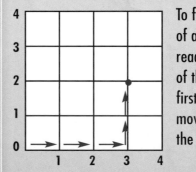

To find the coordinates of a point on a grid, read along the bottom of the grid to find the first coordinate, then move up the grid to find the second coordinate.

Example: Move 3 spaces along the bottom, then 2 spaces up to find coordinates (3, 2) for the point on the grid above.

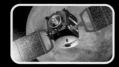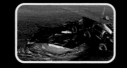

PAGES 14–15

Math Challenge

Military time, shown on the graph on page 15, does not use "a.m." or "p.m." Instead, time is written from 00:00 to 24:00, where both 00:00 and 24:00 are midnight and 12:00 is noon. Any time less than 12:00 falls in the morning (a.m.), and any time greater than 12:00 falls in the afternoon (p.m.)

PAGES 18–19

Mission File

To multiply a number by 100, move the digits two places to the left, and use zeros to fill any empty columns.

Example: 2.8 x 100 = 280

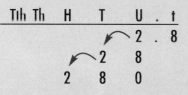

Tth	Th	H	T	U	.	t
				2	.	8
			2	8		
		2	8	0		

Math Challenge

1 miles = 1,760 yards

To change miles per hour into yards per hour, multiply by 1,760.

To change yards per hour into yards per minute, divide by 60.

To change yards per minute into yards per second, divide by 60.

Remember to round up numbers ending in 5, 6, 7, 8 or 9 to the next whole number and round down numbers ending in 1, 2, 3, 4 to the previous whole number.

(25.67 rounded to one decimal place is 25.7).

PAGES 20–21

Mission File

TOP TIP: Percent (%) is a special form of a fraction. *Percent* means "part of 100," so 50% means $^{50}/_{100}$.

A fraction in its simplest form cannot be made with any smaller whole numbers on the top or the bottom. The fraction $^{50}/_{100}$ in its simplest form is ½.

PAGES 24–25

Mission File

To show probability, you can draw a line like the one below and mark intervals, or equal spaces, on the line. Then make Xs, or crosses, on the line to mark the probabilities. The X on this line shows the probability of an event that is very likely to happen.

(very likely)

|———|———|———|———|———X———|
impossible   even chance   certain

PAGES 26–27

Math Challenge

Remember, one year has twelve months. To change a number of years into months, multiply the number of years by 12.

To multiply by 12, first multiply the number by 10, then double the answer, and add the two answers together.

Example: 47 x 12 =

47 x 10	= 470
double 47 (47 x 2)	= 94
470 + 94	= 564
so 47 x 12	= 564

ANSWERS

PAGES 6-7

Mission File

1) a) 37
   b) 322

2) a) 1,852 miles
   b) 1,853 miles
   c) 1,108 miles

3) c (3 times)

Math Challenge

1) Mercury, Venus, Earth, Mars, Jupiter, Saturn, Uranus, Neptune, Pluto
2) Venus
3) a) 342 million miles   c) 2,653 million miles
   b) 745 million miles

PAGES 8-9

Mission File

1) a) October 18, 2009   2) a) April 18, 2010
   b) November 25, 2011   b) May 25, 2012

Math Challenge

122.5   122.6   122.7   122.8   122.9   123

PAGES 10-11

Mission File

1) a) 05:45
   b) 10:15
   c) 14:25
   d) 16:35
   e) 17:38
   f) 17:41
   g) 17:44
   h) 17:44 and 56 seconds
   i) 17:44 and 57 seconds

2) a) 30, 20, 10   d) 9, 6, 3
   b) 8, 6, 4,   e) 9, 5, 1
   c) 12, 7, 2   f) 32, 21, 10

Math Challenge

1) 10 miles   5) 1,500 miles
2) 50 miles   6) 3,000 miles
3) 150 miles   7) 9,000 miles
4) 300 miles   8) 18,000 miles

PAGES 12-13

Mission File

a) 17,600 mph   e) −9
b) 16,200 mph   f) −6
c) 0.3 pounds   g) 24
d) 3

Math Challenge

1) (7, 2) (7, 4) (8, 6)
2) (2, 4)
3) (3, 3)
4) (5, 5)

PAGES 14-15

Mission File

1) 5 °F
2) −44 °F
3) a) −22 °F   c) −22 °F
   b) 50 °F   d) −42 °F

Math Challenge

1) True   4) False — it's about 15 °F
2) True   5) True
3) True   6) False — it's about 04:00

PAGES 16-17

Mission File

1) a) August   2) a) 47.8 °F
   b) January   b) 61.2 °F
   c) November   c) 64.2 °F
   d) May and August
   e) June   3) a) 1.1
   f) February   b) 1.5
   g) May and October   c) 2

Math Challenge

−60.6 °F

Mission File

1) 260 pounds
2) 140 pounds
3) 110 pounds
4) 90 pounds
5) 40 pounds
6) 30 pounds
7) 10 pounds
8) 28,000 pounds
9) 20 pounds

Math Challenge

1) a) 17,600 yards per hour
   b) 293.3 yards per minute
   c) 4.9 yards per second

2) a) 105,600 yards per hour
   b) 1,760 yards per minute
   c) 29.3 yards per second

3) a) 264,000 yards per hour
   b) 4,400 yards per minute
   c) 73.3 yards per second

4) a) 880,000 yards per hour
   b) 14,666.7 yards per minute
   c) 244.4 yards per second

5) a) 30,624,000 yards per hour
   b) 510,400 yards per minute
   c) 8,506.7 yards per second

Mission File

1) 44 hours
2) 25%
3) a) $\frac{1}{5}$
   b) $\frac{3}{10}$
   c) $\frac{1}{4}$

Math Challenge

1) 06:10   07:40   09:10
2) 11:25   12:55   14:25

Mission File

1) 33, 41, 49, 57
2) 35, 43, 51, 59
3) 38, 46, 54
4) 40, 48, 56

Math Challenge

October 27, November 4, November 12, and November 20

Mission File

| 1 | 5 | 4 and 7 | 3 | 2 | 8 | 6 |

impossible   even chance   certain

Math Challenge

1) 17
2) 13
3) Site 3
4) 9
5) 48

Mission File

Amy	31	could be an astronaut
Bethany	26	too young to be an astronaut
Calvin	29	could be an astronaut
Donna	38	could be an astronaut
Emily	41	too old to be an astronaut
Fred	57	too old to be an astronaut

Math Challenge

1) 69 years
2) 828 months

GLOSSARY

ASTRONAUTS people who are specially trained for space travel

ATMOSPHERE the layer of gases, or air, surrounding a planet

COORDINATES the points on a grid where two lines meet or intersect

CORE RADIUS the distance from the center of a planet to the edge of its core, or inner layer

COSMIC RAYS streams of very high energy particles traveling through space at close to the speed of light

DEBRIS the pieces or remains of something broken or destroyed; garbage

DEHYDRATED containing less water than normal

DETERIORATE to weaken a condition or quality; to steadily grow worse

EQUATORIAL RADIUS the distance from the center of a planet's core to the planet's equator, which is an imaginary line around the middle of the planet

EROSION the wearing away of rocks or soil by wind, water, or ice

ESCAPE VELOCITY the minimum speed at which a rocket must be traveling to get away from the pull of Earth's gravity

GEOLOGY the study of a planet's physical features and material structure, such as its rocks and minerals

GRAVITY a force that attracts masses toward each other and that draws objects toward the center of a planet

GROUND CONTROL a team of people on Earth that sends commands to spacecrafts and communicates with astronauts in space

INTERNATIONAL SPACE STATION (ISS) a research station in space, which orbits Earth. The ISS is run by space agencies from the United States, Europe, Canada, Japan, and Russia.

MEAN a number that best represents the middle value of a set of numbers. To find the mean of a set of numbers, you find the total value of the set, then divide by the amount of numbers in the set.

MONITOR to check, watch, or keep track of, usually for a special purpose

ORBIT (n) the circular path an object takes as it moves around another object; (v) to move in a circular path around an object

OXIDIZED combined with oxygen

POLAR RADIUS the distance from the center of a planet's core to the planet's north or south pole

PROBE an unmanned spacecraft that collects information about objects in space and sends the information back to scientists on Earth

REVOLVE to move in a circular path around a center point

SATELLITES objects that orbit a planet or a moon

SOLAR PANELS panels that have special cells, which change energy that comes from sunlight into electricity

SOLAR SYSTEM the Sun and the group of planets and other objects that revolve around it

Measurement Conversions

1 inch = 2.54 centimeters (cm)

1 foot = 0.3048 meter (m)

1 mile = 1.609 kilometers (km)

1 pound = 0.4536 kilograms (kg)

$°F = (°C \times 1.8) + 32$